FAR ISLANDS

FAR ISLANDS

A Documentary Account in Black and White
of FAR Islands in Micronesia Over 30 Years

by

PETER BANG

Peter Bang: FAR ISLANDS
A Documentary Account in Black and White
of FAR Islands in Micronesia Over 30 Years.
© Copyright by Peter Bang 2018, Copenhagen, Denmark
/ Remote Frontlines – *www.remotefrontlines.com*
English translation/edit by Amanda Hansen Bøllehus

This edition is published in black & white on the basis of the color edition of the book "FAR ISLANDS - A Photographer´s Eyewitness Account of FAR Islands in Micronesia Over 30 Years" by Peter Bang.

Publisher: Books on Demand GmbH, Copenhagen, Denmark
Manufacturing: Books on Demand GmbH, Norderstedt, Germany
The book is produced on-Demand-process
ISBN 978-87-430-0368-7

Photos:
NASA: p. 107. *Jovan & AJ Haleyalpiy:* p. 140-141.
Photos showing the author are taken by various friends.
Other photos, cover, map and layout by Peter Bang.

To my Uwapei family

Hosa gashigeshig ngali Refeshailap

All rights reserved.
Photographic, mechanical, or other reproduction of this book or parts there of is prohibited without the author´s permission. Exceptions are brief quotes for use in reviews.

Being strong is only like many sections of bamboo.
Strength does not depend on ones size.

                                        Micronesian proverb

A people's culture resides in the hearts and in the soul of its people. But if you do not tell the stories - if you do not sing the songs and if you do not speak the language – the culture will cease to exist. Our islands, our culture, our traditions, our language, our food, our history and our values are vital to uniting us as a people, it's the foundations upon which we build our identity. People without knowledge of their past, origin and culture is like a tree without root. Our children are tomorrow carriers of our cultural heritage. What we teach our children becomes a part of their identity. Teachers who love teaching, teach children to love learning.

Inspired by spoken words of Uwapei elders and cultural teachers

FAR Islands ~ Faraulep Atoll

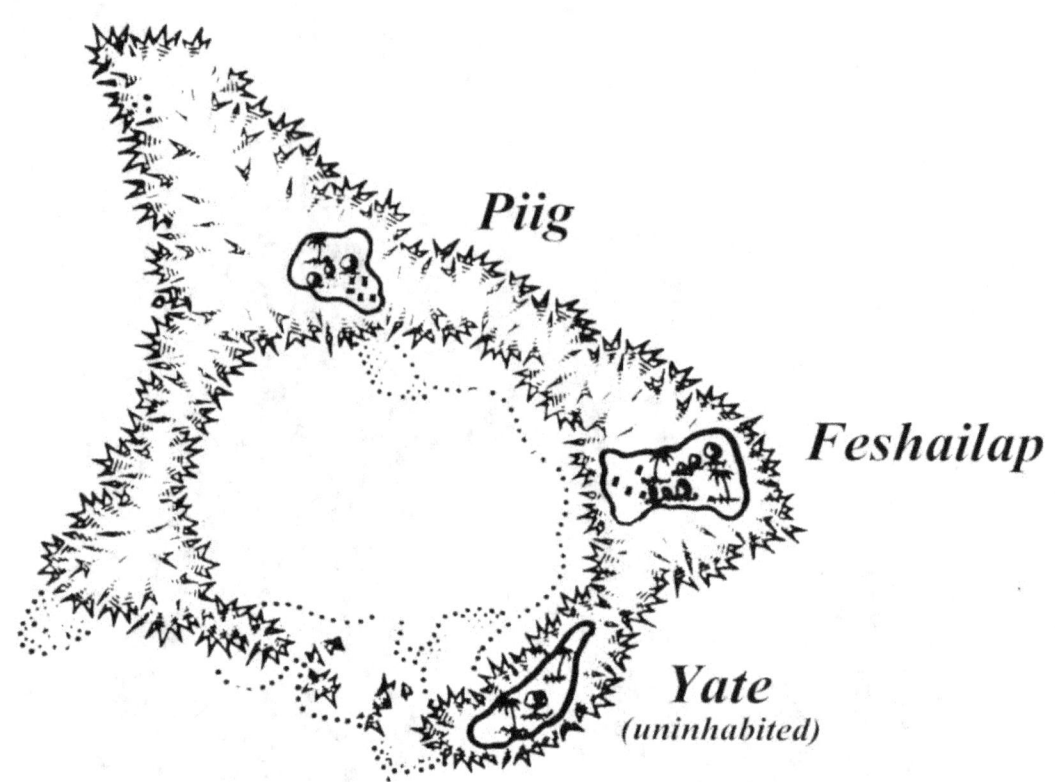

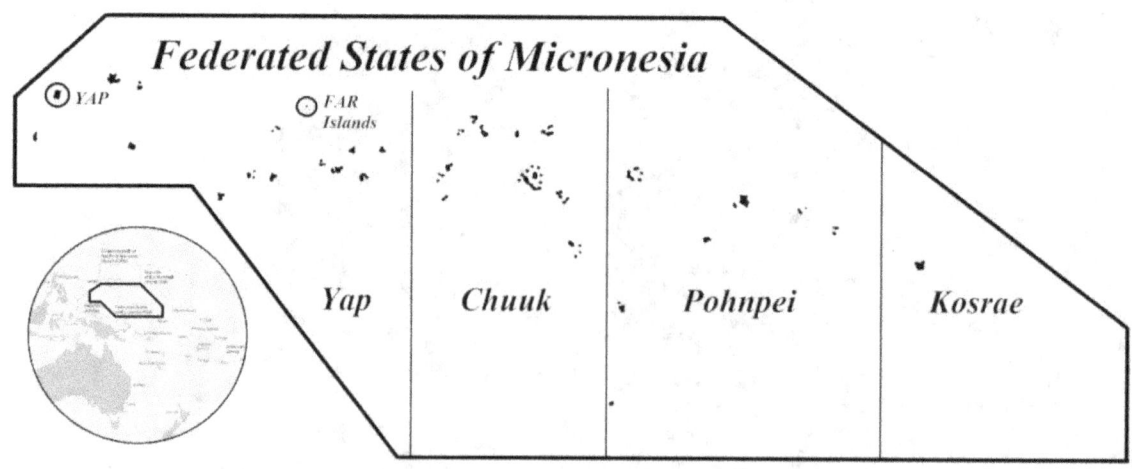

Contents

Map 10
Back then 12
Three decades later 24
Yap 26
The sea voyage 33
Outer islands 40
FAR Island 48
Disappearing islands 56
The Birds Island 80
The last coral garden 92
Typhoon 97
A lone frigatebird 125
State of emergency 136
Bibliography 146

Back then ...

When I first visited Faraulep Atoll three decades ago people on the outer islands in the State of Yap lived much like their ancestors had done for centuries. Contact with the outside world was limited. It was difficult and took time to get there. The only official connection was a ship that belonged to the small island state and served the outer islands only about once every two months.

    Back then, FAR Islands was probably the closest one could get to a paradise on Earth. The white sandy beaches were the epitome of a tropical island dream; the islands were covered with coconut palms and other trees that covered the island's interior and made it a wild and dense jungle.

I was adopted into Pius Mopiy's, who was the island's best fisherman and very helpful and hospitable. In addition to my adoptive father Pius, the family comprised of my mother Meggi and grandmother Mariana and the kids Angie, Pius, Cypriano, Vicky, Rophei and Tom & Jerry.

Below: Family portrait from 1986. From left to right: Mariana (grandmother), Angie, Pius, Meggi (mother), the little boy standing in front of Meggi is Jerry, then Tom. Above Tom is Cypriano. At the far right Vicky and Pius Mopiy (father). Rophei, baby of 6 months, sleeping in the hut (not pictured).

Right: Daily life in the village on Feshailap.

The people of FAR Islands lived a sustainable life and had a rich culture. Their ancestors had survived on the tiny isolated atoll for centuries. They were people of the open sea and depended completely on the ocean. They were great fishermen, and experts in canoe building and navigating the ocean over long distances with their traditional outrigger canoes.
Page 18-19: Turtle party.

I spent a few months on Faraulep Atoll. After returning home, I often missed the islands. But time passed and many waves washed ashore.

It wasn´t until decades later I got in touch with my adoptive brother Pius, who had moved to Hawaii and found me on the Internet. After that I got in contact with others who had been children and young people while I was on the island, but who were now grown and had moved away. Among them were a couple of my sisters and my brother Cypriano. They now lived on the island of Guam, which is a US territory.

Top and right: Beach on Feshailap. Page 22-23: Rush hour in Guam.

Three decades later

After some years of contact via the Internet several of my adoptive family members urged me to go back to the atoll, where a part of the family still lived. I just had to arrange for flights to Micronesia and they would organize the rest. After three decades of absence, I returned to Micronesia, where I met with my adoptive brother Cypriano in Guam. He had not been home on his native island for 15 years, since his work and finances did not allow vacation for more than 2-3 weeks at a time.

Photo below: Reunion with my adoptive brother Cypriano in Guam after three decades.

In Guam lived around 45 people of the Uwapei tribe from FAR Islands. Among them my little sister Rophei, who was a baby of six months the last time I saw her. My sister Angie, who back then was a young girl of 14-15 years, lived there too. Now she was 30 years older and had adult children who had never been on the island and did not speak the native language of Woleaian, but only American. In addition, I met uncles, aunts and other family. They had arranged welcome and farewell parties with lovely food, speeches, music and dance.

Photo below: Uwapei sisters and nieces in Guam capture a selfie before the welcome and farewell party. They were born many years after the author stayed on FAR Islands.

Yap

After a few days in Guam I flew on to Yap, which measures roughly 6 x 15 miles and is the main island in Yap State, the westernmost of the four small island states in the Federated States of Micronesia. The State of Yap has a population of about 19,000 people, of whom approximately 12,000 live on the main island of Yap and the rest on the outer islands. Micronesia, covering a sea area of 2.9 million square miles in the western Pacific, includes several other small island states. The total land area in Micronesia is only about 1,500 square miles spread over 2,100 islands, of which only around 100 are inhabited.

Also on Yap was a great party where among others the appointed chief of Piig, the smallest of the three islands in the atoll, gave a speech. Afterwards he invited me to stay on Piig where I had not lived before, so of course I accepted the invitation.

The feast on Yap took place in a large gathering cottage near the harbor of the small capital of Colonia. People from the outer islands can be party people. And here there was a party. There was also a traditional dance where I was to dance with chief Sugura of Feshailap Island.

Right: Chief Soumai of Piig Island speaking at the welcome party in Yap.
Photo page 28-29: Uwapei dance during the party in Yap.

Daily life on Yap was not all fun and games. The people from the outer islands have no rights to land on the main island of Yap, as the islanders here consider themselves high caste people and the people from the outer islands low caste people who have no right to own land on Yap.

There are two small communities in Yap, where some 50 people from Faraulep Atoll are allowed to stay. Seen through privileged western eyes it was a slum. I saw a huge contrast to my experiences 30 years earlier. There had been tremendous changes; it was a shock to me to see my family of the small islands living in these conditions. I did not stay at a hotel but lived with the Uwapei tribe and experienced staying in several places on Yap together with people from Faraulep Atoll.

Uwapei homes in Yap.

The sea voyage

I stayed on Yap for 10 days before the ship departed. There is only one ship servicing the outer islands. The fieldtrip goes on about every two months and is the only official connection to the remote outer islands. The ship carried 169 passengers who, apart from myself, were all native islanders. There were no cabins for passengers. All passengers slept on the deck where we stayed with our luggage.

Left: The ship serving the outer islands in State of Yap. Below: Seamen.

People laid on the deck like canned sardines at night. We were a group of 16 people going to Faraulep Atoll, including the two chiefs and their families from the atoll's two inhabited islands. Some nights it was difficult to fall asleep, especially in windy and rainy weather when the ship was tossing, so both passengers and luggage moved from side to side while children were seasick and wept.

Below, right and page 36-37: Passengers on the deck.

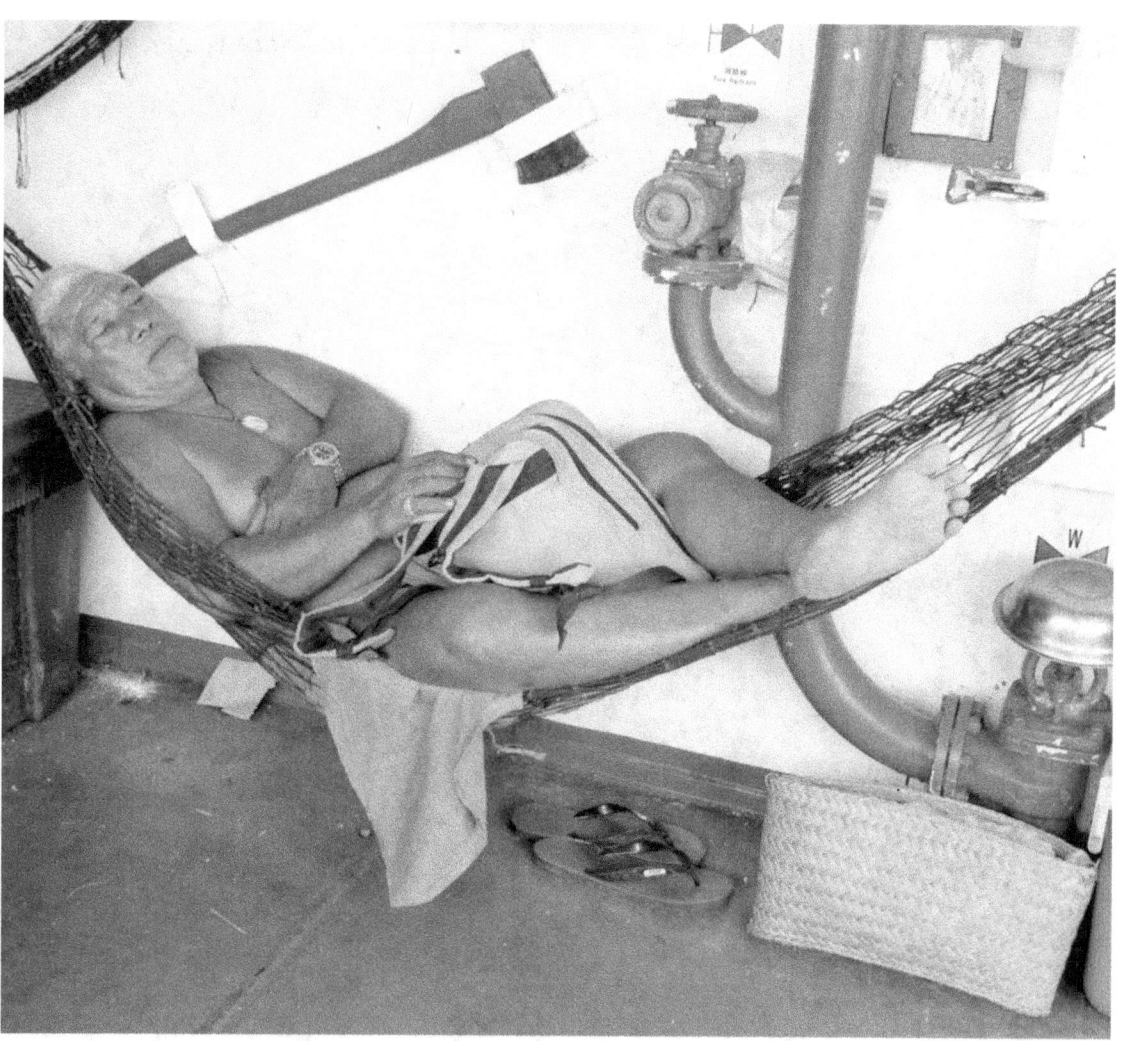

Outer islands

We visited all the inhabited outer islands along the way. Usually we anchored one day each place. Some places we stayed only a few hours.

Page 38-39: Eauripik Atoll, one of the tiny outer islands in State of Yap.

Below and right: Arrival to Ifalik.

41

The ship brought different supplies. But also a great opportunity to meet friends and relatives, and hear the latest news from other islands. Some had been away for many years due to education or work and were welcomed with wreaths of flowers, party and tuba; the local palm wine.

Stress is almost unknown on the outer islands, which is probably one of the most peaceful places on earth.

Daily life on some of the remote islands was in many ways still traditional and the men house an important gathering place for the men and of great social and cultural value.

*Left: Traditional drill.
Right: Fish trap.
Below: Making a fish trap.*

FAR Island

An early morning after a week at sea we had the three small islands in Faraulep Atoll in sight. They did not look like much. The islands highest point is only 12 feet above sea level.

The measure of Piig, the smallest island, is 450 x 450 yard. During my first visit in 1986 only 40 people stayed on Piig, but now the population was 96. Feshailap, the largest island is 800 yard long and 450 yard wide. Before, 135 people lived there and now the population had dropped to 120. The southernmost island, Yate, measures 750 x 250 yard and is only inhabited by rats, crabs, birds and ghosts.

Below: FAR Islands - the three small islands in Faraulep Atoll seen from the ship upon arrival in the early morning. Far left is Piig Island, in the middle Feshailap Island and to the right Yate Island.

Returning after three decades was like waking up after a trip in a time machine. Colossal changes had happened. My brother Jerry, who during my first stay was a little fellow of 5-6 years old, came out and met me on the ship. Now he was over six feet tall, a teacher and employee at the island's school. Jerry still lived with the family, which had long ago abandoned the small palm leaf hut and moved into a concrete house with a tin roof. My adoptive father Pius Mopiy had died four years earlier and was buried in front of the little house. My mother Meggi lived in the house, as did grandma Mariana who with her over 90 years, was the island's oldest. In the house also lived grandchildren and other family members.

Returning home.

Family Photo taken on Feshailap three decades after the authors first visit. The palm leaf hut is replaced with a house of concrete with tin roof. Three of the people in the old image of 1986 (page 14) were still living together in the family house on the island: Grandma Mariana (sitting with a cane and in a white T-shirt), Meggi (mother, standing in the doorway) and Jerry standing to the right of Mariana with his little daughter in his arms. The others in the picture are kids, grandchildren and cousins.

Right: The authors adoptive mother Meggi cooking with grandchildren.

The family lived centrally in the village next to the church, which compared to earlier days was a huge edifice. Previously the church was a palm-leaf hut that looked like the other huts, just a bit bigger.

The islanders became Catholics after World War II when a Catholic missionary went ashore. The islanders rapidly converted, as their traditional God Uwapei via a spiritual media some months earlier had announced that there would soon be a new administration, which used a cross as a symbol and that Uwapei then no longer should be their God. Uwapei who was worshiped as a guardian angel, was a spirit of a little girl who, after her death had been a good adviser because all she had done in her short life had been good and full of love for others. Uwapei spoke with opposite words, so all she said meant the opposite. Uwapei could suddenly appear as a bird or a fish at sea. But she could always be found in a rainbow.

Electricity was another big change. During my first stay the nearest location with electricity was in Yap, 435 miles away. Now, with the support of European Union, solar cells had been installed on all the inhabited outer islands.

In front of the school a satellite dish had been set up for receiving and sending wave mail, but it was broken and had been out of order for 5 years. The only contact with the outside world remained a small VHF radio, which was in the office at the school that was managed by Tom, my second brother on the island, and the island's school principal. My brothers Tom and Jerry were both trained as teachers and had returned to teach the children on their native island.

Right: Portrait of Uwapei girl wearing flower wreath.

Youth on Feshailap.

Disappearing islands

Although there were many changes the hospitality and sound of laughing children was the same. Children live in the moment and know only what they see and experience this exact moment. The children I saw running and playing in the shallow water near the beach could not see the effects of climate change. They had never seen the coral reef when it was alive or the beach in front of the men´s house before erosion really took off. Their future would be very different from the lives their parents lived. Tom believed that it is all about educating the children to a future outside the islands as they are disappearing.

Below: Daily life on Feshailap. Children outside their home. Right: Boy sitting on the stump of a coconut tree that once grew on the ground.

FAR Islands is on the frontline of climate change. Every day the islanders witnessed the ongoing erosion caused by the rising sea water.

The photo above was taken at low tide showing the erosion along the coastline. The photo to the left was taken during high tide. Page 60-61: Erosion, fallen coconut trees on the beach, Feshailap. Page 62-63: Woman of the Uwapei tribe walking in the water where there used to be land before.

Global warming was a recurring topic of conversation in the men's house. Everyone was concerned about climate change. The melting of the ice sheet in Greenland over the past two decades resulted in an average sea level rise of 0,6 inches per year in the Western Pacific near the Equator, so over 20 years the sea has risen one foot.

One of the things that struck me most was to see how things were at the coral reef. I almost had a shock when I saw all the dead corals at low tide. And when I went swimming out to my old favorite place in the lagoon, where before I enjoyed going snorkeling, I became so sad, that moisture appeared in the mask from the inside, so the glass began to fog. When last I saw the reef it was a wonderful underwater garden full of living creatures of all shapes and colors. Now it was completely dead.

Left top: Conversation in front of the men's house on Feshailap.
Left below: Fallen coconut trees and eroded coast on Feshailap due to rising sea water caused by climate change.
Below: View over the dead coral reef at low tide.

Above: A healthy and living coral reef at low tide on FAR Atoll in 1986.

The same view at low tide three decades later... All the corals are dead.

The corals and the algae it lived in symbiosis with, had not been able to withstain the record high temperatures. The reef had started to die about 15 years earlier, which had severely affected all life in the coral reef. Many species had disappeared, died or gone elsewhere. The old ecosystem of the coral reef was completely collapsed, so almost no fish were left in the lagoon.

When I snorkeled it happened that I would see a fish trap lying on the bottom. Fish traps were the only fishing gear that was really used in the lagoon. The chief had forbidden the use of harpoons and fishing spears to protect the population of fish, just as he had banned fishing with nets. The only fishing gear that was otherwise used in the lagoon were fish hooks with fishing line, and a short metal rod bent at the end like a hook used for catching clams, lobsters and octopus.

Bottom and right: Fish trap.

Right: Getting the fish trap. Top and bottom: Fish caught in the fish trap.

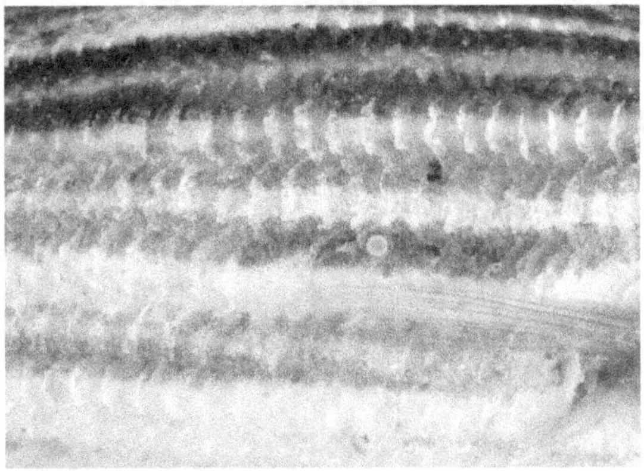

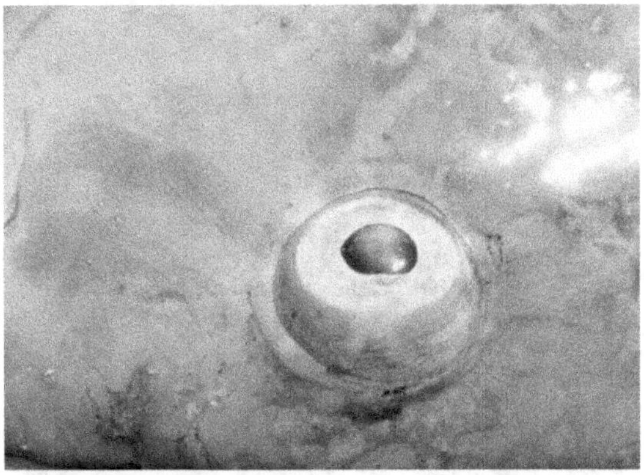

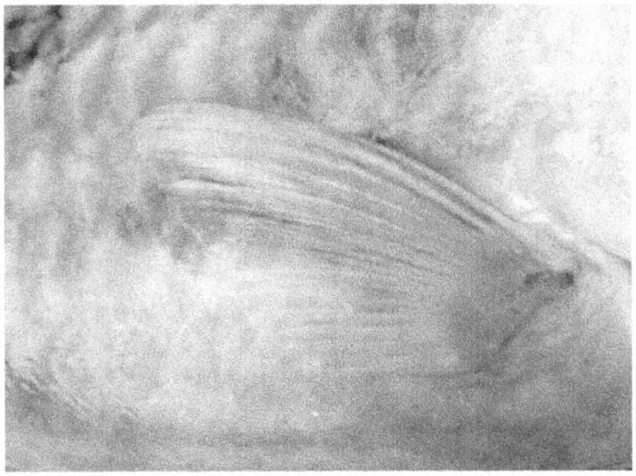

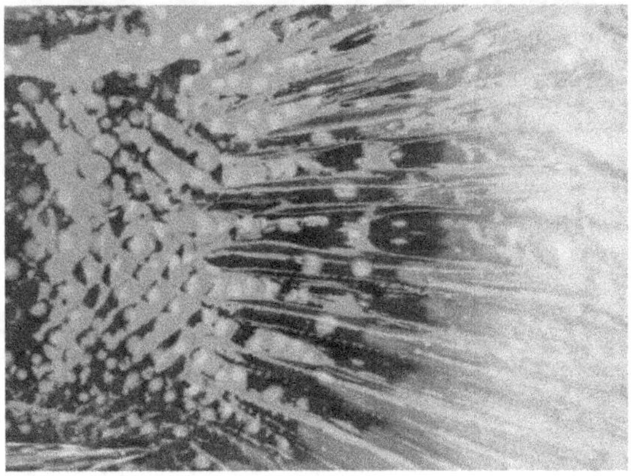

Most families had their own fish traps, which were inspected about every other day. Sometimes the fish traps were empty, while other days full enough for everyone in the family to get a small fish. Usually they went out 3-4 men together emptying their traps from a fiberglass boat with an outboard engine.

Left: Jerry shares the catch.

Below: Time to eat.

The islands were one hundred percent self-sufficient when it came to food. The women cultivated the gardens on the island, where there grew taro, bananas, sweet potatoes, pumpkin, papaya and other crops.

Left: Returning from garden work.

Below: Planting sweet potatoes.

Left: Women prepare taro for dinner.

Right: Time to sleep.

Below: Village life.

The Birds Island

After three weeks on Feshailap my brother Jerry sailed me to the small island Pigg, as during the party on Yap the chief had invited me to visit, so I could experience the atoll's smallest island.

On Piig I lived in a small hut, located for itself near the island's northernmost spit of land, which was uninhabited. When I moved into the hut it was leaking and a ramshackle. It was only used as a storage place for coconut shells, used as firewood for cooking. But chief Soumai and his family helped restore the hut, so rain and wind were kept reasonably out.

Photo below and right: The authors hut on Piig Island.
Page 82-83: The natural environment around the hut on Piig.

For two months the author lived in the hut side by side with crabs, geckos and lizards.

Top: Lizard.

Left: Coconut crab.

Right: Stranded coconut ready to sprout on Piig.

In the woods and on the beach there were lots of birds, including migratory birds that like I were on winter visit from far northern distant skies, where it was now cold and dark. Piig, or Pigue Maal, which is the island's ancient indigenous name, means the birds' Island.

Photo below: Writing diary inside the hut.

Right: Birds outside the hut / top left: White tern. Top right: Black noddy terns and a flying frigatebird. Middle left: Whimbrel. Middle right: Grey reef heron. Bottom: Turnstone.

However, the birds were not my only neighbors. Chief Soumai and his family lived in the forest about 100 meters behind my hut. They took care of me in every way. The hospitality was tremendous. Each morning Soumai came and sat down in the opening of my hut, where we started the day with the taste of coconut juice and a chat about the weather or what else there was to talk about on a tiny island, which at times was quite a lot.

Soumai was by far the most traditional and active fisherman on the island. If the weather permitted it he paddled outside the lagoon every evening before sunset. Here he would cast a coral block as anchor and sit down to fish. Usually he returned with his catch after midnight.

Soumai ready for night fishing.

The big catches were done with fiberglass dinghy and outboard motor. Sometimes it was quite big yellow fin tuna that were caught. When that happened, there was food. As there was not yet freezer on the atoll the catch could not be stored, so the people on the islands lived from hand to mouth and never caught more than the population could eat during a day or two.

Left: Wahoo - caught on hand line on the open ocean.

Below: Fresh tuna.

The last coral garden

On the reef in the lagoon west of Piig was a coral garden that was protected from the warm sea currents from the east, which had made the rest of the coral reef collapse. In a part of this area the chief had prohibited fishing, so it served as a reserve chamber having a reasonable fish stock and living corals, where I enjoyed snorkeling.

Another place I liked to dwell was on some coral rocks that lay out towards a place with strong currents. Off the coast the water was 12-16 feet deep with a sandy bottom and scattered coral blocks. It was a great place to fish so occasionally I could contribute a little to supper.

Typhoon

After six weeks on Piig suddenly something happened that changed everything. It was announced over the radio that a strong typhoon was coming. It had been named Maysak and was heading for the islands.

The rest of the day and the next day were spent securing the island's roofs with rope that at suitable intervals were tied to the socket, stumps and bottom of the traditional huts´ supporting pillars which were dug one meter into the ground. The ropes were then fastened tightly above the roof and fixed on the other side.

Left: Soumai blows a conch shell to show how his ancestors tried to chase the evil sea spirit of a typhoon away.

Below: Bad weather is coming

Page 98-99: Huge waves were crashing towards the reef.

The next morning we woke up to heavy winds. Along with Soumai and Nicky, I stood on the beach and saw the waves thundered towards the reef. I am still haunted by the expression in Soumai's eyes when the first giant wave broke into the reef and rolled further up above the beach, so Nicky quickly escaped and my hut was flooded. When we went back to school we heard many heavy bumps of coconuts that fell to the ground. It did not feel safe outside.

Below, right and page 100-101: Flooding around the authors hut.

We were about 200 people on the atoll, who were in danger of being washed into the sea if it was high tide when the typhoon hit. In the office at the school we barricaded ourselves before midnight. Suddenly the door was beaten up with an explosive wind and the latch ripped out of the bracket, so we had to block the door from the inside with a large plastic barrel three quarters full of fresh water, which we pushed in front of the flapping door while the light bulb in the ceiling flashed and the children cried. This was followed by prayers to God.

While the typhoon was raging at night the island's population stayed in the small school that served as typhoon shelter. The pictures were taken a few hours before the typhoon hit with full power.

We didn't consider that somewhere up in the sky one of NASA satellites hovered recording images of the typhoon's eye seen from above, where we were just below. The route of the typhoon was carefully recorded and mapped. Large international TV stations published NASA's images and news updates from the few places in Micronesia that could be reported from with pictures of Super Typhoon Maysak's rampage. But nothing was told about the small islands in the center of the eye.

The typhoon arrived at approximately 8 pm. But it wasn't until one hour after midnight, at approximately 1 am that the center of the typhoon hit with full power and wind speed of 140 miles per hour. Before that, the radio connection went off and the antenna was blown away, so we had no connection to the outside world.

Outside it sounded as if the world was about to go down. The noise was so hellish that it drowned the sound of the crying children. There was a crazy orchestra of noisy torrential sounds from all sides. The wind shrieked and howled through the narrow cracks around the clacking metal shutters. Large broken branches were crashing against the walls of the house outside. Later, we were told there had been gusts up to 160 miles per hour. Every now and then we heard the sound of big trees whose trunks were broken and either splintered or toppled and uprooted.

Right: Typhoon Maysak strengthened into a super typhoon on March 31, 2015, the night the typhoon hit FAR Atoll, reaching Category 5 hurricane status on the Saffir-Simpson Wind Scale. These satellite photos were taken by NASA and shows Typhoon Maysak.

The portraits in the following two pages 108-109 shows over 200 people of the Uwapei tribe, equal to the total population of FAR Atoll at the time the typhoon hit the islands.

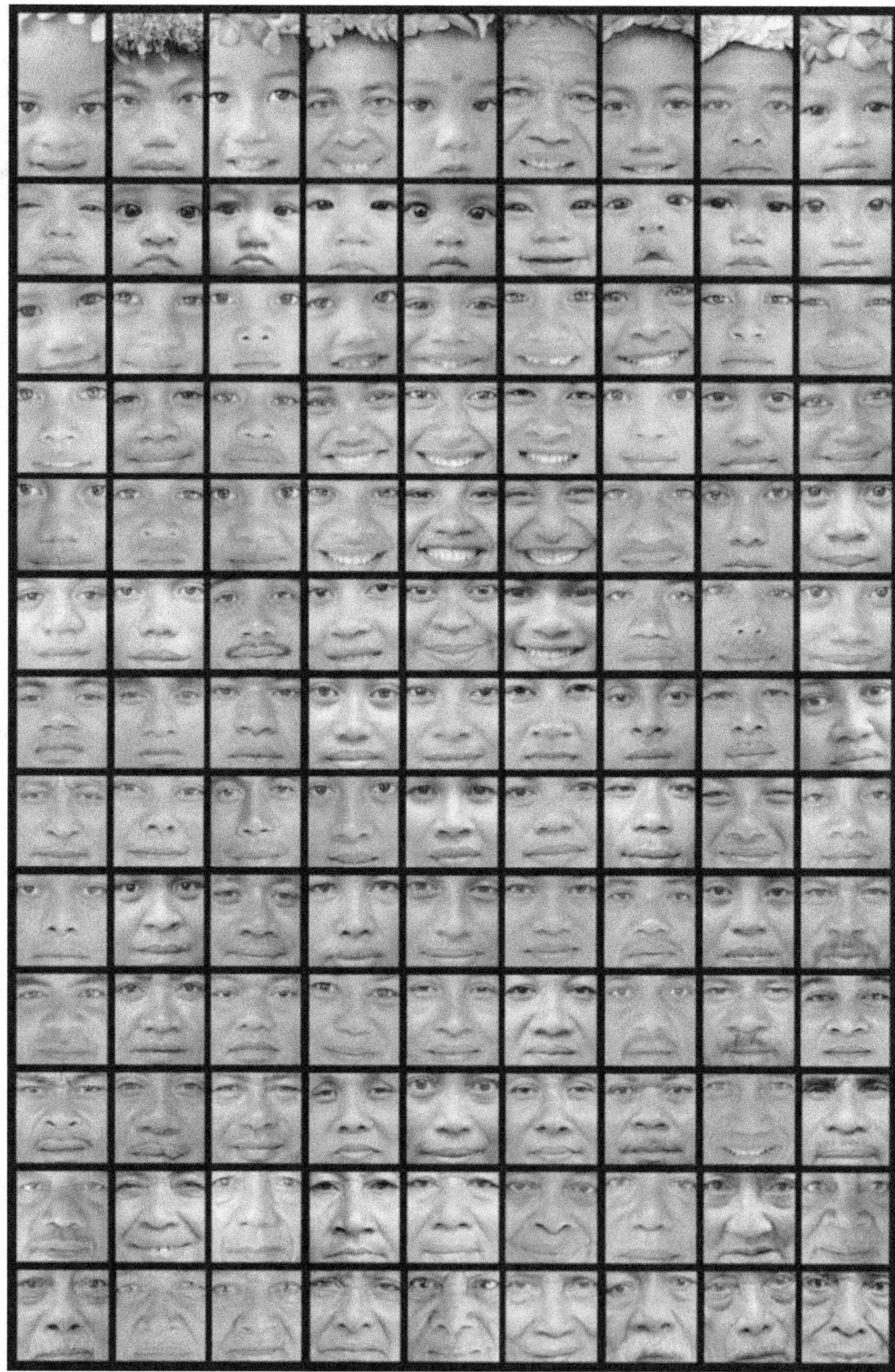

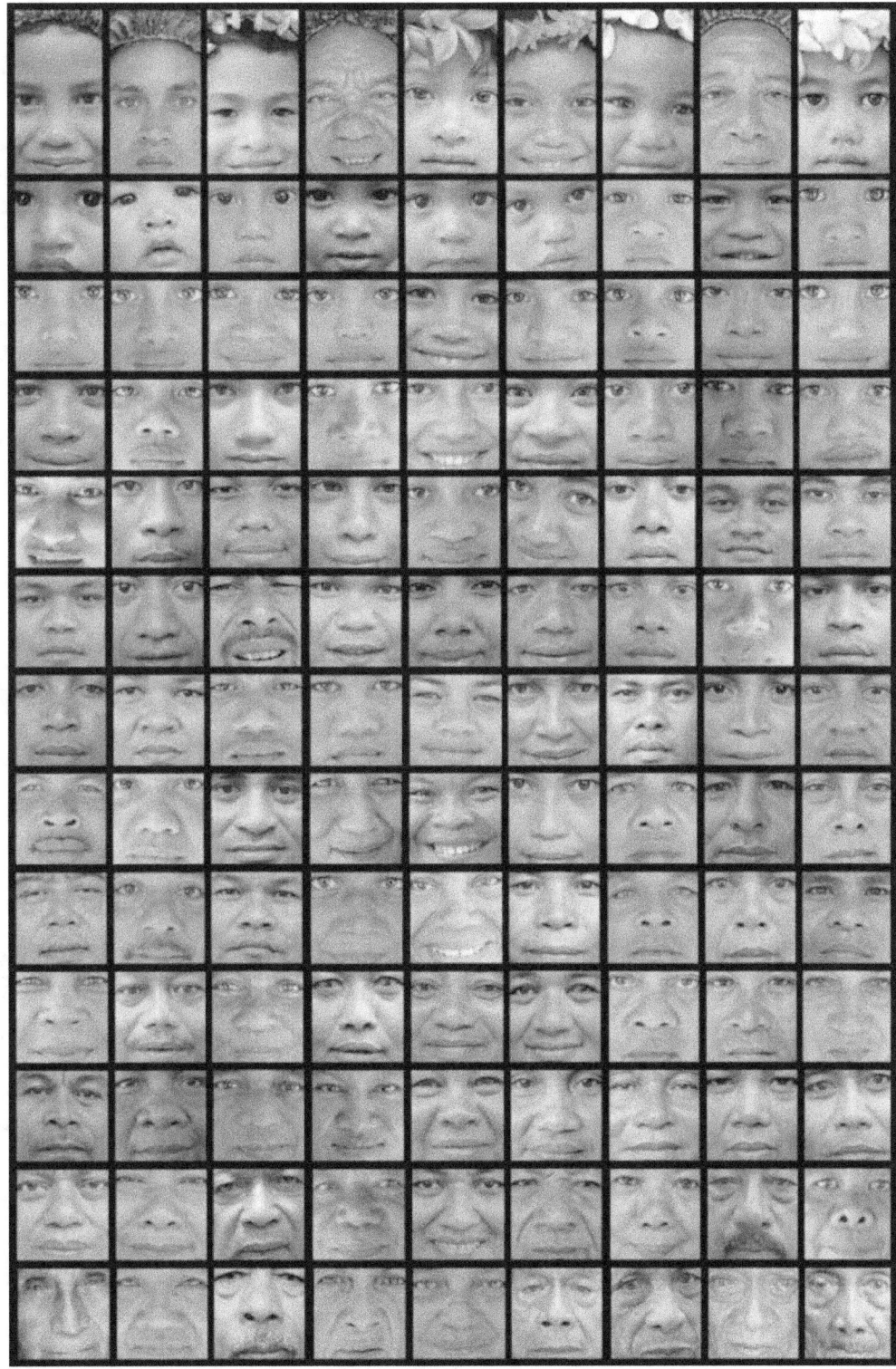

When daylight came many wanted to go outside in the storm and rain to see how extensive the damage was. Soumai´s granddaughter Alona of 2 ½ years, who stayed in the same room as me, had like everyone else, not slept much, so it was a tired little girl who this morning looked out on a world that during the night had changed dramatically.

While watching the kids I realized that they might have children of their own before it will again be possible to find shade where the trees now laid toppled and broken.

Above: Alona gets her first glimpse of the island the day after the typhoon.

Right: Fallen trees on Piig Island the day after Typhoon Maysak.

There were 96 people on the island who had to get something to eat. All around lay injured birds, especially terns and their fallen chicks that had been blown out of nests. Consequently the morning's meal came to consist of wild sea birds roasted over a fire behind the school and consumed with coconuts and a bit of freshly cooked wild taro root, dug up in the forest.

Left: Kids climb a broken tree.

Below: A sparse breakfast the morning after the typhoon. A grilled tern on a plate; one of the many injured seabirds that were consumed by hungry people the days after the typhoon.

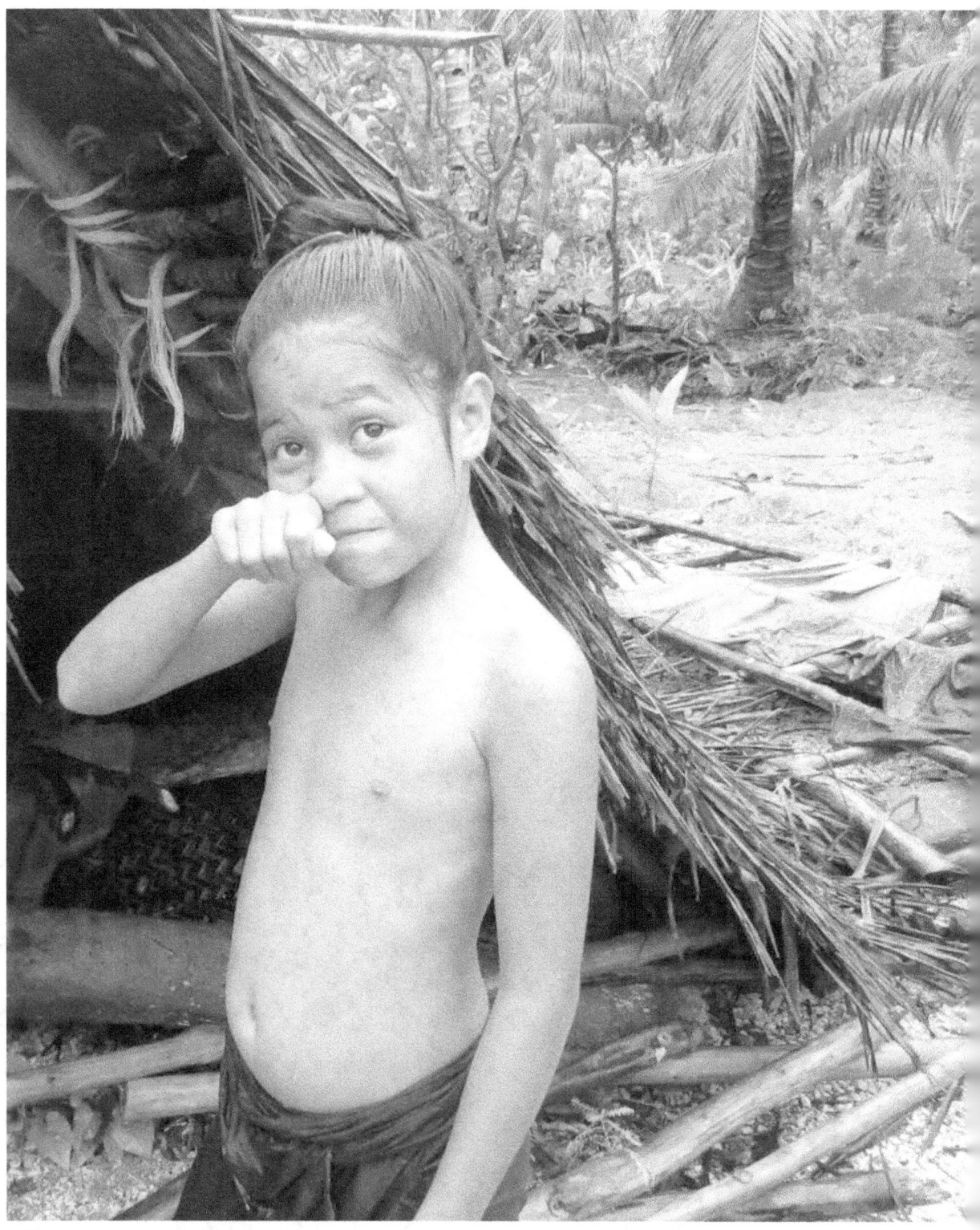

Considering the power of the strong winds the typhoon had brought, I was surprised there were still huts left on the island. Some were of course blown away; the roof blown off, sides and ends blown in or they had been destroyed by fallen trees. But most remained thanks to ancient construction and securing with rope over the roof. Considering the circumstances my own hut had done well, although the sides were flushed away, the gable pressed and the roof awry.

Left, top and page 116-117: A sad and shocking moment to see everything that one knows and loves being so totally destroyed and changed.

Some places it was utter chaos and impossible to get around. Almost all the big breadfruit trees were down, branches and treetops blown to pieces, leaving only the trunks with long wood splinters bristling out in all directions. Papaya and banana trees were nearly all broken, overturned or torn up by the roots and leaves blown into a thousand pieces.

There were fallen coconuts everywhere, which after a few days turned bad. Many coconut trees were toppled and broken. A palm tree I walked by in the forest had been pushed so hard by the wind, that the middle of the trunk was split lengthwise and twisted so the tree was bent down with the crown weighed down to the ground. The coconut tree´s long, strong roots and resilient strain caused the roots were still firmly entrenched in the ground so that the bowed trunk formed an arch, which stood as a portal amidst the windswept palm forest.

It was worse where the waves in many places had destroyed the coast and eaten several yards inland making the island even smaller, so there was now sandy beach and fallen coconut trees helter-skelter, where before there was land.

Left and top: On a walk across the island the author found this bowed coconut tree in the devastated forest where the typhoon had left a trail of chaos.

Top, below and right: The beach around the island the day after the typhoon. Page 122-123: A boy of the Uwapei tribe standing his ground on a fallen coconut tree in strong wind on Piig Island the day after Typhoon Maysak.

A lone frigatebird

Fresh water wells were flooded by seawater and some of them covered by a thick layer of sand so that the well had disappeared. The taro patches was swamped by one meter of salt water and completely destroyed. As salt water seeped out of the fields, the dead inedible crops remained. The soil was destroyed by salt and the land made impossible to cultivate for several years.

A few days after the typhoon, the tide was extremely low and the weather very hot, windless, dry and sunny. The reef had drained so one could walk between the islands. I went out on a part of the reef, where I had been before at low tide. Prior to the typhoon this field was covered with dead corals. At high tide octopus were often hunted here. Now it was completely changed and without a single hiding place for octopus. All the dead coral blocks were swept away by the waves and the strong current.

Left: A temporary surviving coconut tree stands on a new beach where it was dry land the day before. Above: Frigatebird.

Top: Flooded taro patch on Piig Island the day after Typhoon Maysak.
Below: Dying taro in the same area a week after the typhoon.

Above: Dead coral blocks on the reef at low tide, this photo was taken a few weeks before the typhoon hit. Below: The same area after the typhoon.

In the afternoon, after the typhoon hit at night, the radio antenna came up again and contact to the outside world was established. Yet it was not possible to get in touch with Feshailap, so the next day two boats were put in the water and along with a third of Piig's population I went to Feshailap.

Here a breadfruit tree had fallen into the school and destroyed the roof. Luckily no one got hurt. As I walked past the church a lone frigate bird flew out of the sky over Gabriel and Peter as they took up the cross to naile it to the top of the gable, from where it had been blown down.

Below: Two boats with a third of Piig's population went to Feshailap.

Right top: The school on Feshailap was hit by a breadfruit tree.

Right bottom: The cross on the island's church was blown down. A lone frigate bird flew out of the sky above Feshailap as Gabriel and Peter were about to nail the cross firmly to the gable of the church. Uwapei, the traditional God on the atoll could suddenly appear as a frigatebird.

A huge clean-up was waiting. Thousands of things had to be repaired and taken care of. The boats on Feshailap had been carried far up on land and could not be put in the water before the access to the beach was cleared for fallen trees. Hand axes and machetes were the only tools, so it would take several days before the boats could get into water.

In the afternoon a big meeting was held at the men's house with the participation of all the men. The actual cleaning was planned in order to work in teams around the church and the school.

During my stay many islanders had talked a great deal about climate change so during the meeting I suggested we made a protest which I would try to photograph and record on video and then try to get published as a call for help to the outside world, which there was wide support for.

The demonstration was quickly arranged and started in the water in front of the men's house, which had previously been dry land. Then the procession went to the school and further through a tangle of fallen trees to the church and back to the beach where the protest ended.

Below, right and page 132-133: Protest on Faraulep Atoll.

The time that followed was difficult. After a couple of very rainy days came a long period of drought, so we really had to conserve the drinking water that was unhealthy, even if it was boiled. It was almost depressing to stay on the island. Before, the environment was green, but now the island was wilted and brown from dead leaves. And everywhere day and night the sound of screaming and cawing of homeless birds was heard, calling their dead chicks that lay stinking in hundreds under leaves, trees and downed branches, and which caused a hell of flies on the islands.

It was pure misery. Millions of flies crawling everywhere, which resulted in all of us becoming sick and getting diarrhea, vomiting, headaches, fever and abdominal pain.

Top: Two weeks after the typhoon.

Right: Hair cut after the typhoon.

State of emergency

Fourteen days after the typhoon I fell down from a big box that tipped because I put myself on top of it to hang some gear to dry under the eaves. I fell into a tailspin and crashed onto the edge of the box with the left side of my chest so that two ribs were pressed in and broken in three places.

The following four days I was treated with local massage and medicine in the form of chopped leaves and herbs mixed with coconut oil and rubbed on the left side of my chest, while the mattress I was lying on was covered with a thick layer of chopped green leaves. At the same time, I started a powerful penicillin cure against infection in internal bleedings.

The massage began early morning and spanned over four sessions a day, each session about two hours, a total of around 8 hours of intensive massage per day. This was done without anesthetic so sometimes I had to be fixated. But already after two days of treatment it was amazing to feel how effective the massage and the whole treatment were.

When I later went to the hospital and was x-rayed and scanned so the doctors could see the fractures they were impressed by how well the bones were connected and put in the right place. No modern surgeon could have done it better by surgery, rather the contrary, I was told.

The first three days I couldn´t do anything. I was in the same position on the back all the time and needed help with everything. My body was constantly soaked in oil and chopped green leaves and herbs. In the daytime, when I did not get massaged women sat around me with palm leaf fans to cool me and keep flies away.

Right top: Traditional medicine was made of many different leaves, herbs, coconut oil and more.

Right bottom: After the accident the author was healed with local massage and medicine.

A cargo ship from the central government brought emergency relief in the form of medicine, bags of rice, boxes of canned food and clean bottled drinking water.

Six days after the accident, a cargo ship from the central government on the island of Pohnpei brought relief in the form of medicine, bags of rice, boxes of canned food and clean bottled drinking water.

After three months on Faraulep Atoll I was sailed directly to Yap as a patient. When the ship arrived an ambulance was waiting on the dock, ready to take me to the island's hospital. Despite the circumstances the journey home was a joyful series of welcome and farewell dinners as well as supporting protests calling the outside world for help. Before departure from Guam a public page on Facebook, called Uwapei Global Warming Campaign was also created.

Three weeks after the accident I sat on an airplane high above the Pacific Ocean.

Some months later, the State of Yap was declared to be in a state of emergency due to the extreme heat and drought that lasted for almost six month, and which meant that the last crops and the coral garden west of Piig died.

Less than a year after the typhoon, about 10% of the islands population had moved and lived with their family on Yap, Guam and other islands in the Western Pacific and other places, where several nations of low-lying islands and atolls are expected to disappear and be depopulated within the next 20-30 years due to sea level rise caused by global warming.

But when this account was written most of the people of the Uwapei tribe still lived on FAR Islands. A year after the typhoon, a man from the islands sent me a picture of a wide range of dancing men wearing palm leaf skirts, traditional body paint and flower wreaths in rainbow colors as they sang and danced Uwapei's old dance to celebrate the new year, the rain's return and the end of the long draught, so the island now again is green.

Photo page 140-141: Men of the Uwapei tribe celebrate New Year on Feshailap by singing and dancing the old traditional Uwapei dance. The picture was taken on Feshailap by Jovan with AJ Haleyalpiy's camera (AJ is dancing in the middle of the picture / number four from left).

"How long will our islands survive? How will we survive as a people, as a culture?"

Sam Ilesugam
(Uwapei Guam)

Bibliography

Al Gore: An Inconvenient Truth - The Crisis of Global Warming (2006). Former US Vice President Al Gore's New York Times bestselling book is a daring call to action, exposing the shocking reality of how humankind has aided in the destruction of our planet and the future we face if we do not take action to stop global warming.

William H. Alkire: Lamotrek Atoll And Inter-Island Socioeconomic Ties. Waveland Press, Inc., USA 1965.

William H. Alkire: Coral Islanders. University of Victoria, British Columbia, Canada 1978.

Gene Ashby: Micronesian Custums And Beliefs. Published by the Education Department, Trust Territory of the Pacific Islands 1975.

Peter Bang: The Last Turtle Party - Endangered Native People in Micronesia (updated and published in English 2018). The original edition "Den sidste skildpaddefest" was first published in 1991 in Danish by Borgen Publishing, Copenhagen, Denmark.

Internet

Facebook:
Disappearing Islands: Info and updates related to FAR Islands and this book.
Uwapei Global Warming Campaign: Public international page telling about the impact of climate change on FAR Atoll and other Pacific islands.
Remote Frontlines: Info and updates about isolated and forgotten endangered indigenous people and natural environment and wildlife around the globe; Pacific, Asia, Africa, Greenland, the north Canadian wilderness and other places.

More info:

www.remotefrontlines.com

By the same author:

Peter Bang: PAPUA BLOOD
A Photographer´s Eyewitness Account of West Papua Over 30 Years

Over an interval spanning three decades the author and photographer Peter Bang describes his experiences among the indigenous people of West Papua, who are threatened by a continuing history of genocide and extinction.

Published 2018 by Remote Frontlines.
248 pages. 198 color photos.

Peter Bang: PUWUL´S WORLD
Endagered Native People

A book for children telling about a boy who live in a Stone Age culture in the central highlands of West Papua and experience his first contact with the outside world.

Published 2018 by Remote Frontlines.
48 pages. 50 color photos.

Peter Bang: THE LAST TURTLE PARTY
Endagered Native People in Micronesia

"The Last Turtle Party" is a tale for children about the boy Mau who, together with his father, travels in a canoe to the neighboring island to participate in the annual turtle party.

Published 2018 by Remote Frontlines.
48 pages. 50 color photos.

Peter Bang

Born in Denmark 1957. Lecturer and author of half a dozen books published in Danish, including travel accounts and photo documentary books about indigenous people. In addition to numerous articles, photographic exhibitions and publications, including publications for the United Nations International Year of Indigenous People 1993. Travel and subsistence among indigenous people
in West Papua, Africa, Asia, North America and Greenland. Former member of FIJET, the International Federation of Journalists and Travel Writers.

www.ingramcontent.com/pod-product-compliance
Lightning Source LLC
Chambersburg PA
CBHW082205220526
45470CB00010B/3052